M000198228

3D Printing Logbook

Learn from 3D Printing Failures and Ensure Continuous Improvement in Print Quality, Maintenance and Speed through Systematic Record Keeping

TABLE OF CONTENTS

Copyrighted
DISCLAIMER

HOW TO USE THIS LOGBOOK

Thank you for purchasing this logbook. There are 4 different sections to this book, and I will go through each of them. Each section has several unfilled tables; along with 1 filled table. You can use that filled table as an example if you are confused about how to fill in the table after the print.

a. **Individual Print Log:** This is what you can fill after each individual print. This will give you a good record of what settings produce good/bad results; and you can catch on if there is any deterioration of print quality or speed over time.

b. **Failure Log:** This is a space to record all failures, their root causes and their fixes. This has helped me out immensely. There were a couple of issues that happened recently with my nozzle. I had no idea how to fix it; but I noticed in my failure log that it had happened two years earlier and the root cause was the wrong nozzle used for flexible filament. It saved me at least 10 hours of work.

c. **R&D/Testing Log:** This is a great resource if you want to test the impact of certain variables (temperature, filament material etc.) on the outputs (print speed, print quality, weight etc.). Great for anyone doing R&D or any other sort of testing.

d. **Maintenance Log:** This is a great place to record any changes or addition of parts to the 3D Printer; and also schedule any future prints.

Individual Print LOG

Print Number _

Inputs	
Printer Name	
Print File Name	
Filament Material	
Bed Temperature	
Nozzle Temperature	
Nozzle Diameter	
Layer Height	
Initial Layer Height	
Flow	
Enable Retraction (Yes/No)	
Wall Thickness	
Travel Speed	
Print Acceleration	
Travel Acceleration	
Print Jerk	
Travel Jerk	
Build Plate Adhesion Type	
Level of Support Structures	
Comments….	
Outputs	
Print Completion Time	
Smoothness Level (1-10)	
Weight	
Volume	
Noise Level	

(Example) Print Number _

Inputs	
Printer Name	Ultimaker
Print File Name	pencilholder.stl
Filament Material	PLA
Bed Temperature	50 C
Nozzle Temperature	250 C
Nozzle Diameter	12 mm
Layer Height	0.3 mm
Initial Layer Height	0.1 mm
Flow	100%
Enable Retraction (Yes/No)	No
Wall Thickness	0.1 mm
Travel Speed	50 mm/sec
Print Acceleration	500 mm/sec^2
Travel Acceleration	5000 mm/sec^2
Print Jerk	20 mm/sec
Travel Jerk	30 mm/sec
Build Plate Adhesion Type	Raft
Level of Support Structures	No support Structures
Comments....	Test Print
Outputs	
Print Completion Time	1 hour 43 minutes
Smoothness Level (1-10)	8 (Estimation)
Weight	120 grams
Volume	180 cm^3
Noise Level	20 dbA

Print Number _

Inputs	
Printer Name	
Print File Name	
Filament Material	
Bed Temperature	
Nozzle Temperature	
Nozzle Diameter	
Layer Height	
Initial Layer Height	
Flow	
Enable Retraction (Yes/No)	
Wall Thickness	
Travel Speed	
Print Acceleration	
Travel Acceleration	
Print Jerk	
Travel Jerk	
Build Plate Adhesion Type	
Level of Support Structures	
Comments....	
Outputs	
Print Completion Time	
Smoothness Level (1-10)	
Weight	
Volume	
Noise Level	

Print Number _

Inputs	
Printer Name	
Print File Name	
Filament Material	
Bed Temperature	
Nozzle Temperature	
Nozzle Diameter	
Layer Height	
Initial Layer Height	
Flow	
Enable Retraction (Yes/No)	
Wall Thickness	
Travel Speed	
Print Acceleration	
Travel Acceleration	
Print Jerk	
Travel Jerk	
Build Plate Adhesion Type	
Level of Support Structures	
Comments….	
Outputs	
Print Completion Time	
Smoothness Level (1-10)	
Weight	
Volume	
Noise Level	

Print Number _

Inputs	
Printer Name	
Print File Name	
Filament Material	
Bed Temperature	
Nozzle Temperature	
Nozzle Diameter	
Layer Height	
Initial Layer Height	
Flow	
Enable Retraction (Yes/No)	
Wall Thickness	
Travel Speed	
Print Acceleration	
Travel Acceleration	
Print Jerk	
Travel Jerk	
Build Plate Adhesion Type	
Level of Support Structures	
Comments….	
Outputs	
Print Completion Time	
Smoothness Level (1-10)	
Weight	
Volume	
Noise Level	

Print Number _

Inputs	
Printer Name	
Print File Name	
Filament Material	
Bed Temperature	
Nozzle Temperature	
Nozzle Diameter	
Layer Height	
Initial Layer Height	
Flow	
Enable Retraction (Yes/No)	
Wall Thickness	
Travel Speed	
Print Acceleration	
Travel Acceleration	
Print Jerk	
Travel Jerk	
Build Plate Adhesion Type	
Level of Support Structures	
Comments….	
Outputs	
Print Completion Time	
Smoothness Level (1-10)	
Weight	
Volume	
Noise Level	

Print Number _

Inputs	
Printer Name	
Print File Name	
Filament Material	
Bed Temperature	
Nozzle Temperature	
Nozzle Diameter	
Layer Height	
Initial Layer Height	
Flow	
Enable Retraction (Yes/No)	
Wall Thickness	
Travel Speed	
Print Acceleration	
Travel Acceleration	
Print Jerk	
Travel Jerk	
Build Plate Adhesion Type	
Level of Support Structures	
Comments….	
Outputs	
Print Completion Time	
Smoothness Level (1-10)	
Weight	
Volume	
Noise Level	

Print Number _

Inputs	
Printer Name	
Print File Name	
Filament Material	
Bed Temperature	
Nozzle Temperature	
Nozzle Diameter	
Layer Height	
Initial Layer Height	
Flow	
Enable Retraction (Yes/No)	
Wall Thickness	
Travel Speed	
Print Acceleration	
Travel Acceleration	
Print Jerk	
Travel Jerk	
Build Plate Adhesion Type	
Level of Support Structures	
Comments….	
Outputs	
Print Completion Time	
Smoothness Level (1-10)	
Weight	
Volume	
Noise Level	

Print Number _

Inputs	
Printer Name	
Print File Name	
Filament Material	
Bed Temperature	
Nozzle Temperature	
Nozzle Diameter	
Layer Height	
Initial Layer Height	
Flow	
Enable Retraction (Yes/No)	
Wall Thickness	
Travel Speed	
Print Acceleration	
Travel Acceleration	
Print Jerk	
Travel Jerk	
Build Plate Adhesion Type	
Level of Support Structures	
Comments....	
Outputs	
Print Completion Time	
Smoothness Level (1-10)	
Weight	
Volume	
Noise Level	

Print Number _

Inputs	
Printer Name	
Print File Name	
Filament Material	
Bed Temperature	
Nozzle Temperature	
Nozzle Diameter	
Layer Height	
Initial Layer Height	
Flow	
Enable Retraction (Yes/No)	
Wall Thickness	
Travel Speed	
Print Acceleration	
Travel Acceleration	
Print Jerk	
Travel Jerk	
Build Plate Adhesion Type	
Level of Support Structures	
Comments….	
Outputs	
Print Completion Time	
Smoothness Level (1-10)	
Weight	
Volume	
Noise Level	

Print Number _

Inputs	
Printer Name	
Print File Name	
Filament Material	
Bed Temperature	
Nozzle Temperature	
Nozzle Diameter	
Layer Height	
Initial Layer Height	
Flow	
Enable Retraction (Yes/No)	
Wall Thickness	
Travel Speed	
Print Acceleration	
Travel Acceleration	
Print Jerk	
Travel Jerk	
Build Plate Adhesion Type	
Level of Support Structures	
Comments....	
Outputs	
Print Completion Time	
Smoothness Level (1-10)	
Weight	
Volume	
Noise Level	

Print Number _

Inputs	
Printer Name	
Print File Name	
Filament Material	
Bed Temperature	
Nozzle Temperature	
Nozzle Diameter	
Layer Height	
Initial Layer Height	
Flow	
Enable Retraction (Yes/No)	
Wall Thickness	
Travel Speed	
Print Acceleration	
Travel Acceleration	
Print Jerk	
Travel Jerk	
Build Plate Adhesion Type	
Level of Support Structures	
Comments….	
Outputs	
Print Completion Time	
Smoothness Level (1-10)	
Weight	
Volume	
Noise Level	

Print Number _

Inputs	
Printer Name	
Print File Name	
Filament Material	
Bed Temperature	
Nozzle Temperature	
Nozzle Diameter	
Layer Height	
Initial Layer Height	
Flow	
Enable Retraction (Yes/No)	
Wall Thickness	
Travel Speed	
Print Acceleration	
Travel Acceleration	
Print Jerk	
Travel Jerk	
Build Plate Adhesion Type	
Level of Support Structures	
Comments....	
Outputs	
Print Completion Time	
Smoothness Level (1-10)	
Weight	
Volume	
Noise Level	

Print Number _

Inputs	
Printer Name	
Print File Name	
Filament Material	
Bed Temperature	
Nozzle Temperature	
Nozzle Diameter	
Layer Height	
Initial Layer Height	
Flow	
Enable Retraction (Yes/No)	
Wall Thickness	
Travel Speed	
Print Acceleration	
Travel Acceleration	
Print Jerk	
Travel Jerk	
Build Plate Adhesion Type	
Level of Support Structures	
Comments....	
Outputs	
Print Completion Time	
Smoothness Level (1-10)	
Weight	
Volume	
Noise Level	

Print Number _

Inputs	
Printer Name	
Print File Name	
Filament Material	
Bed Temperature	
Nozzle Temperature	
Nozzle Diameter	
Layer Height	
Initial Layer Height	
Flow	
Enable Retraction (Yes/No)	
Wall Thickness	
Travel Speed	
Print Acceleration	
Travel Acceleration	
Print Jerk	
Travel Jerk	
Build Plate Adhesion Type	
Level of Support Structures	
Comments….	
Outputs	
Print Completion Time	
Smoothness Level (1-10)	
Weight	
Volume	
Noise Level	

Print Number _

Inputs	
Printer Name	
Print File Name	
Filament Material	
Bed Temperature	
Nozzle Temperature	
Nozzle Diameter	
Layer Height	
Initial Layer Height	
Flow	
Enable Retraction (Yes/No)	
Wall Thickness	
Travel Speed	
Print Acceleration	
Travel Acceleration	
Print Jerk	
Travel Jerk	
Build Plate Adhesion Type	
Level of Support Structures	
Comments….	
Outputs	
Print Completion Time	
Smoothness Level (1-10)	
Weight	
Volume	
Noise Level	

Print Number _

Inputs	
Printer Name	
Print File Name	
Filament Material	
Bed Temperature	
Nozzle Temperature	
Nozzle Diameter	
Layer Height	
Initial Layer Height	
Flow	
Enable Retraction (Yes/No)	
Wall Thickness	
Travel Speed	
Print Acceleration	
Travel Acceleration	
Print Jerk	
Travel Jerk	
Build Plate Adhesion Type	
Level of Support Structures	
Comments....	
Outputs	
Print Completion Time	
Smoothness Level (1-10)	
Weight	
Volume	
Noise Level	

Print Number _

Inputs	
Printer Name	
Print File Name	
Filament Material	
Bed Temperature	
Nozzle Temperature	
Nozzle Diameter	
Layer Height	
Initial Layer Height	
Flow	
Enable Retraction (Yes/No)	
Wall Thickness	
Travel Speed	
Print Acceleration	
Travel Acceleration	
Print Jerk	
Travel Jerk	
Build Plate Adhesion Type	
Level of Support Structures	
Comments….	
Outputs	
Print Completion Time	
Smoothness Level (1-10)	
Weight	
Volume	
Noise Level	

Print Number _

Inputs	
Printer Name	
Print File Name	
Filament Material	
Bed Temperature	
Nozzle Temperature	
Nozzle Diameter	
Layer Height	
Initial Layer Height	
Flow	
Enable Retraction (Yes/No)	
Wall Thickness	
Travel Speed	
Print Acceleration	
Travel Acceleration	
Print Jerk	
Travel Jerk	
Build Plate Adhesion Type	
Level of Support Structures	
Comments….	
Outputs	
Print Completion Time	
Smoothness Level (1-10)	
Weight	
Volume	
Noise Level	

Print Number _

Inputs	
Printer Name	
Print File Name	
Filament Material	
Bed Temperature	
Nozzle Temperature	
Nozzle Diameter	
Layer Height	
Initial Layer Height	
Flow	
Enable Retraction (Yes/No)	
Wall Thickness	
Travel Speed	
Print Acceleration	
Travel Acceleration	
Print Jerk	
Travel Jerk	
Build Plate Adhesion Type	
Level of Support Structures	
Comments….	
Outputs	
Print Completion Time	
Smoothness Level (1-10)	
Weight	
Volume	
Noise Level	

Copyrighted
Print Number _

Inputs	
Printer Name	
Print File Name	
Filament Material	
Bed Temperature	
Nozzle Temperature	
Nozzle Diameter	
Layer Height	
Initial Layer Height	
Flow	
Enable Retraction (Yes/No)	
Wall Thickness	
Travel Speed	
Print Acceleration	
Travel Acceleration	
Print Jerk	
Travel Jerk	
Build Plate Adhesion Type	
Level of Support Structures	
Comments....	
Outputs	
Print Completion Time	
Smoothness Level (1-10)	
Weight	
Volume	
Noise Level	

Print Number _

Inputs	
Printer Name	
Print File Name	
Filament Material	
Bed Temperature	
Nozzle Temperature	
Nozzle Diameter	
Layer Height	
Initial Layer Height	
Flow	
Enable Retraction (Yes/No)	
Wall Thickness	
Travel Speed	
Print Acceleration	
Travel Acceleration	
Print Jerk	
Travel Jerk	
Build Plate Adhesion Type	
Level of Support Structures	
Comments....	
Outputs	
Print Completion Time	
Smoothness Level (1-10)	
Weight	
Volume	
Noise Level	

Print Number _

Inputs	
Printer Name	
Print File Name	
Filament Material	
Bed Temperature	
Nozzle Temperature	
Nozzle Diameter	
Layer Height	
Initial Layer Height	
Flow	
Enable Retraction (Yes/No)	
Wall Thickness	
Travel Speed	
Print Acceleration	
Travel Acceleration	
Print Jerk	
Travel Jerk	
Build Plate Adhesion Type	
Level of Support Structures	
Comments....	
Outputs	
Print Completion Time	
Smoothness Level (1-10)	
Weight	
Volume	
Noise Level	

Print Number _

Inputs	
Printer Name	
Print File Name	
Filament Material	
Bed Temperature	
Nozzle Temperature	
Nozzle Diameter	
Layer Height	
Initial Layer Height	
Flow	
Enable Retraction (Yes/No)	
Wall Thickness	
Travel Speed	
Print Acceleration	
Travel Acceleration	
Print Jerk	
Travel Jerk	
Build Plate Adhesion Type	
Level of Support Structures	
Comments….	
Outputs	
Print Completion Time	
Smoothness Level (1-10)	
Weight	
Volume	
Noise Level	

Print Number _

Inputs	
Printer Name	
Print File Name	
Filament Material	
Bed Temperature	
Nozzle Temperature	
Nozzle Diameter	
Layer Height	
Initial Layer Height	
Flow	
Enable Retraction (Yes/No)	
Wall Thickness	
Travel Speed	
Print Acceleration	
Travel Acceleration	
Print Jerk	
Travel Jerk	
Build Plate Adhesion Type	
Level of Support Structures	
Comments....	
Outputs	
Print Completion Time	
Smoothness Level (1-10)	
Weight	
Volume	
Noise Level	

Print Number _

Inputs	
Printer Name	
Print File Name	
Filament Material	
Bed Temperature	
Nozzle Temperature	
Nozzle Diameter	
Layer Height	
Initial Layer Height	
Flow	
Enable Retraction (Yes/No)	
Wall Thickness	
Travel Speed	
Print Acceleration	
Travel Acceleration	
Print Jerk	
Travel Jerk	
Build Plate Adhesion Type	
Level of Support Structures	
Comments….	
Outputs	
Print Completion Time	
Smoothness Level (1-10)	
Weight	
Volume	
Noise Level	

Print Number _

Inputs	
Printer Name	
Print File Name	
Filament Material	
Bed Temperature	
Nozzle Temperature	
Nozzle Diameter	
Layer Height	
Initial Layer Height	
Flow	
Enable Retraction (Yes/No)	
Wall Thickness	
Travel Speed	
Print Acceleration	
Travel Acceleration	
Print Jerk	
Travel Jerk	
Build Plate Adhesion Type	
Level of Support Structures	
Comments….	
Outputs	
Print Completion Time	
Smoothness Level (1-10)	
Weight	
Volume	
Noise Level	

Copyrighted
Print Number _

Inputs	
Printer Name	
Print File Name	
Filament Material	
Bed Temperature	
Nozzle Temperature	
Nozzle Diameter	
Layer Height	
Initial Layer Height	
Flow	
Enable Retraction (Yes/No)	
Wall Thickness	
Travel Speed	
Print Acceleration	
Travel Acceleration	
Print Jerk	
Travel Jerk	
Build Plate Adhesion Type	
Level of Support Structures	
Comments....	
Outputs	
Print Completion Time	
Smoothness Level (1-10)	
Weight	
Volume	
Noise Level	

Print Number _

Inputs	
Printer Name	
Print File Name	
Filament Material	
Bed Temperature	
Nozzle Temperature	
Nozzle Diameter	
Layer Height	
Initial Layer Height	
Flow	
Enable Retraction (Yes/No)	
Wall Thickness	
Travel Speed	
Print Acceleration	
Travel Acceleration	
Print Jerk	
Travel Jerk	
Build Plate Adhesion Type	
Level of Support Structures	
Comments....	
Outputs	
Print Completion Time	
Smoothness Level (1-10)	
Weight	
Volume	
Noise Level	

Print Number _

Inputs	
Printer Name	
Print File Name	
Filament Material	
Bed Temperature	
Nozzle Temperature	
Nozzle Diameter	
Layer Height	
Initial Layer Height	
Flow	
Enable Retraction (Yes/No)	
Wall Thickness	
Travel Speed	
Print Acceleration	
Travel Acceleration	
Print Jerk	
Travel Jerk	
Build Plate Adhesion Type	
Level of Support Structures	
Comments….	
Outputs	
Print Completion Time	
Smoothness Level (1-10)	
Weight	
Volume	
Noise Level	

Print Number _

Inputs	
Printer Name	
Print File Name	
Filament Material	
Bed Temperature	
Nozzle Temperature	
Nozzle Diameter	
Layer Height	
Initial Layer Height	
Flow	
Enable Retraction (Yes/No)	
Wall Thickness	
Travel Speed	
Print Acceleration	
Travel Acceleration	
Print Jerk	
Travel Jerk	
Build Plate Adhesion Type	
Level of Support Structures	
Comments….	
Outputs	
Print Completion Time	
Smoothness Level (1-10)	
Weight	
Volume	
Noise Level	

Print Number _

Inputs	
Printer Name	
Print File Name	
Filament Material	
Bed Temperature	
Nozzle Temperature	
Nozzle Diameter	
Layer Height	
Initial Layer Height	
Flow	
Enable Retraction (Yes/No)	
Wall Thickness	
Travel Speed	
Print Acceleration	
Travel Acceleration	
Print Jerk	
Travel Jerk	
Build Plate Adhesion Type	
Level of Support Structures	
Comments....	
Outputs	
Print Completion Time	
Smoothness Level (1-10)	
Weight	
Volume	
Noise Level	

Print Number _

Inputs	
Printer Name	
Print File Name	
Filament Material	
Bed Temperature	
Nozzle Temperature	
Nozzle Diameter	
Layer Height	
Initial Layer Height	
Flow	
Enable Retraction (Yes/No)	
Wall Thickness	
Travel Speed	
Print Acceleration	
Travel Acceleration	
Print Jerk	
Travel Jerk	
Build Plate Adhesion Type	
Level of Support Structures	
Comments….	
Outputs	
Print Completion Time	
Smoothness Level (1-10)	
Weight	
Volume	
Noise Level	

Print Number _

Inputs	
Printer Name	
Print File Name	
Filament Material	
Bed Temperature	
Nozzle Temperature	
Nozzle Diameter	
Layer Height	
Initial Layer Height	
Flow	
Enable Retraction (Yes/No)	
Wall Thickness	
Travel Speed	
Print Acceleration	
Travel Acceleration	
Print Jerk	
Travel Jerk	
Build Plate Adhesion Type	
Level of Support Structures	
Comments….	
Outputs	
Print Completion Time	
Smoothness Level (1-10)	
Weight	
Volume	
Noise Level	

Print Number _

Inputs	
Printer Name	
Print File Name	
Filament Material	
Bed Temperature	
Nozzle Temperature	
Nozzle Diameter	
Layer Height	
Initial Layer Height	
Flow	
Enable Retraction (Yes/No)	
Wall Thickness	
Travel Speed	
Print Acceleration	
Travel Acceleration	
Print Jerk	
Travel Jerk	
Build Plate Adhesion Type	
Level of Support Structures	
Comments….	
Outputs	
Print Completion Time	
Smoothness Level (1-10)	
Weight	
Volume	
Noise Level	

Print Number _

Inputs	
Printer Name	
Print File Name	
Filament Material	
Bed Temperature	
Nozzle Temperature	
Nozzle Diameter	
Layer Height	
Initial Layer Height	
Flow	
Enable Retraction (Yes/No)	
Wall Thickness	
Travel Speed	
Print Acceleration	
Travel Acceleration	
Print Jerk	
Travel Jerk	
Build Plate Adhesion Type	
Level of Support Structures	
Comments….	
Outputs	
Print Completion Time	
Smoothness Level (1-10)	
Weight	
Volume	
Noise Level	

Print Number _

Inputs	
Printer Name	
Print File Name	
Filament Material	
Bed Temperature	
Nozzle Temperature	
Nozzle Diameter	
Layer Height	
Initial Layer Height	
Flow	
Enable Retraction (Yes/No)	
Wall Thickness	
Travel Speed	
Print Acceleration	
Travel Acceleration	
Print Jerk	
Travel Jerk	
Build Plate Adhesion Type	
Level of Support Structures	
Comments....	
Outputs	
Print Completion Time	
Smoothness Level (1-10)	
Weight	
Volume	
Noise Level	

Print Number _

Inputs	
Printer Name	
Print File Name	
Filament Material	
Bed Temperature	
Nozzle Temperature	
Nozzle Diameter	
Layer Height	
Initial Layer Height	
Flow	
Enable Retraction (Yes/No)	
Wall Thickness	
Travel Speed	
Print Acceleration	
Travel Acceleration	
Print Jerk	
Travel Jerk	
Build Plate Adhesion Type	
Level of Support Structures	
Comments….	
Outputs	
Print Completion Time	
Smoothness Level (1-10)	
Weight	
Volume	
Noise Level	

Print Number _

Inputs	
Printer Name	
Print File Name	
Filament Material	
Bed Temperature	
Nozzle Temperature	
Nozzle Diameter	
Layer Height	
Initial Layer Height	
Flow	
Enable Retraction (Yes/No)	
Wall Thickness	
Travel Speed	
Print Acceleration	
Travel Acceleration	
Print Jerk	
Travel Jerk	
Build Plate Adhesion Type	
Level of Support Structures	
Comments....	
Outputs	
Print Completion Time	
Smoothness Level (1-10)	
Weight	
Volume	
Noise Level	

Print Number _

Inputs	
Printer Name	
Print File Name	
Filament Material	
Bed Temperature	
Nozzle Temperature	
Nozzle Diameter	
Layer Height	
Initial Layer Height	
Flow	
Enable Retraction (Yes/No)	
Wall Thickness	
Travel Speed	
Print Acceleration	
Travel Acceleration	
Print Jerk	
Travel Jerk	
Build Plate Adhesion Type	
Level of Support Structures	
Comments....	
Outputs	
Print Completion Time	
Smoothness Level (1-10)	
Weight	
Volume	
Noise Level	

Print Number _

Inputs	
Printer Name	
Print File Name	
Filament Material	
Bed Temperature	
Nozzle Temperature	
Nozzle Diameter	
Layer Height	
Initial Layer Height	
Flow	
Enable Retraction (Yes/No)	
Wall Thickness	
Travel Speed	
Print Acceleration	
Travel Acceleration	
Print Jerk	
Travel Jerk	
Build Plate Adhesion Type	
Level of Support Structures	
Comments….	
Outputs	
Print Completion Time	
Smoothness Level (1-10)	
Weight	
Volume	
Noise Level	

Print Number _

Inputs	
Printer Name	
Print File Name	
Filament Material	
Bed Temperature	
Nozzle Temperature	
Nozzle Diameter	
Layer Height	
Initial Layer Height	
Flow	
Enable Retraction (Yes/No)	
Wall Thickness	
Travel Speed	
Print Acceleration	
Travel Acceleration	
Print Jerk	
Travel Jerk	
Build Plate Adhesion Type	
Level of Support Structures	
Comments….	
Outputs	
Print Completion Time	
Smoothness Level (1-10)	
Weight	
Volume	
Noise Level	

Print Number _

Inputs	
Printer Name	
Print File Name	
Filament Material	
Bed Temperature	
Nozzle Temperature	
Nozzle Diameter	
Layer Height	
Initial Layer Height	
Flow	
Enable Retraction (Yes/No)	
Wall Thickness	
Travel Speed	
Print Acceleration	
Travel Acceleration	
Print Jerk	
Travel Jerk	
Build Plate Adhesion Type	
Level of Support Structures	
Comments....	
Outputs	
Print Completion Time	
Smoothness Level (1-10)	
Weight	
Volume	
Noise Level	

Print Number _

Inputs	
Printer Name	
Print File Name	
Filament Material	
Bed Temperature	
Nozzle Temperature	
Nozzle Diameter	
Layer Height	
Initial Layer Height	
Flow	
Enable Retraction (Yes/No)	
Wall Thickness	
Travel Speed	
Print Acceleration	
Travel Acceleration	
Print Jerk	
Travel Jerk	
Build Plate Adhesion Type	
Level of Support Structures	
Comments….	
Outputs	
Print Completion Time	
Smoothness Level (1-10)	
Weight	
Volume	
Noise Level	

Print Number _

Inputs	
Printer Name	
Print File Name	
Filament Material	
Bed Temperature	
Nozzle Temperature	
Nozzle Diameter	
Layer Height	
Initial Layer Height	
Flow	
Enable Retraction (Yes/No)	
Wall Thickness	
Travel Speed	
Print Acceleration	
Travel Acceleration	
Print Jerk	
Travel Jerk	
Build Plate Adhesion Type	
Level of Support Structures	
Comments….	
Outputs	
Print Completion Time	
Smoothness Level (1-10)	
Weight	
Volume	
Noise Level	

Print Number _

Inputs	
Printer Name	
Print File Name	
Filament Material	
Bed Temperature	
Nozzle Temperature	
Nozzle Diameter	
Layer Height	
Initial Layer Height	
Flow	
Enable Retraction (Yes/No)	
Wall Thickness	
Travel Speed	
Print Acceleration	
Travel Acceleration	
Print Jerk	
Travel Jerk	
Build Plate Adhesion Type	
Level of Support Structures	
Comments….	
Outputs	
Print Completion Time	
Smoothness Level (1-10)	
Weight	
Volume	
Noise Level	

Print Number _

Inputs	
Printer Name	
Print File Name	
Filament Material	
Bed Temperature	
Nozzle Temperature	
Nozzle Diameter	
Layer Height	
Initial Layer Height	
Flow	
Enable Retraction (Yes/No)	
Wall Thickness	
Travel Speed	
Print Acceleration	
Travel Acceleration	
Print Jerk	
Travel Jerk	
Build Plate Adhesion Type	
Level of Support Structures	
Comments….	
Outputs	
Print Completion Time	
Smoothness Level (1-10)	
Weight	
Volume	
Noise Level	

Print Number _

Inputs	
Printer Name	
Print File Name	
Filament Material	
Bed Temperature	
Nozzle Temperature	
Nozzle Diameter	
Layer Height	
Initial Layer Height	
Flow	
Enable Retraction (Yes/No)	
Wall Thickness	
Travel Speed	
Print Acceleration	
Travel Acceleration	
Print Jerk	
Travel Jerk	
Build Plate Adhesion Type	
Level of Support Structures	
Comments….	
Outputs	
Print Completion Time	
Smoothness Level (1-10)	
Weight	
Volume	
Noise Level	

Print Number _

Inputs	
Printer Name	
Print File Name	
Filament Material	
Bed Temperature	
Nozzle Temperature	
Nozzle Diameter	
Layer Height	
Initial Layer Height	
Flow	
Enable Retraction (Yes/No)	
Wall Thickness	
Travel Speed	
Print Acceleration	
Travel Acceleration	
Print Jerk	
Travel Jerk	
Build Plate Adhesion Type	
Level of Support Structures	
Comments….	
Outputs	
Print Completion Time	
Smoothness Level (1-10)	
Weight	
Volume	
Noise Level	

Print Number _

Inputs	
Printer Name	
Print File Name	
Filament Material	
Bed Temperature	
Nozzle Temperature	
Nozzle Diameter	
Layer Height	
Initial Layer Height	
Flow	
Enable Retraction (Yes/No)	
Wall Thickness	
Travel Speed	
Print Acceleration	
Travel Acceleration	
Print Jerk	
Travel Jerk	
Build Plate Adhesion Type	
Level of Support Structures	
Comments….	
Outputs	
Print Completion Time	
Smoothness Level (1-10)	
Weight	
Volume	
Noise Level	

Print Number _

Inputs	
Printer Name	
Print File Name	
Filament Material	
Bed Temperature	
Nozzle Temperature	
Nozzle Diameter	
Layer Height	
Initial Layer Height	
Flow	
Enable Retraction (Yes/No)	
Wall Thickness	
Travel Speed	
Print Acceleration	
Travel Acceleration	
Print Jerk	
Travel Jerk	
Build Plate Adhesion Type	
Level of Support Structures	
Comments....	
Outputs	
Print Completion Time	
Smoothness Level (1-10)	
Weight	
Volume	
Noise Level	

Print Number _

Inputs	
Printer Name	
Print File Name	
Filament Material	
Bed Temperature	
Nozzle Temperature	
Nozzle Diameter	
Layer Height	
Initial Layer Height	
Flow	
Enable Retraction (Yes/No)	
Wall Thickness	
Travel Speed	
Print Acceleration	
Travel Acceleration	
Print Jerk	
Travel Jerk	
Build Plate Adhesion Type	
Level of Support Structures	
Comments….	
Outputs	
Print Completion Time	
Smoothness Level (1-10)	
Weight	
Volume	
Noise Level	

Print Number _

Inputs	
Printer Name	
Print File Name	
Filament Material	
Bed Temperature	
Nozzle Temperature	
Nozzle Diameter	
Layer Height	
Initial Layer Height	
Flow	
Enable Retraction (Yes/No)	
Wall Thickness	
Travel Speed	
Print Acceleration	
Travel Acceleration	
Print Jerk	
Travel Jerk	
Build Plate Adhesion Type	
Level of Support Structures	
Comments….	
Outputs	
Print Completion Time	
Smoothness Level (1-10)	
Weight	
Volume	
Noise Level	

Print Number _

Inputs	
Printer Name	
Print File Name	
Filament Material	
Bed Temperature	
Nozzle Temperature	
Nozzle Diameter	
Layer Height	
Initial Layer Height	
Flow	
Enable Retraction (Yes/No)	
Wall Thickness	
Travel Speed	
Print Acceleration	
Travel Acceleration	
Print Jerk	
Travel Jerk	
Build Plate Adhesion Type	
Level of Support Structures	
Comments….	
Outputs	
Print Completion Time	
Smoothness Level (1-10)	
Weight	
Volume	
Noise Level	

Print Number _

Inputs	
Printer Name	
Print File Name	
Filament Material	
Bed Temperature	
Nozzle Temperature	
Nozzle Diameter	
Layer Height	
Initial Layer Height	
Flow	
Enable Retraction (Yes/No)	
Wall Thickness	
Travel Speed	
Print Acceleration	
Travel Acceleration	
Print Jerk	
Travel Jerk	
Build Plate Adhesion Type	
Level of Support Structures	
Comments….	
Outputs	
Print Completion Time	
Smoothness Level (1-10)	
Weight	
Volume	
Noise Level	

Print Number _

Inputs	
Printer Name	
Print File Name	
Filament Material	
Bed Temperature	
Nozzle Temperature	
Nozzle Diameter	
Layer Height	
Initial Layer Height	
Flow	
Enable Retraction (Yes/No)	
Wall Thickness	
Travel Speed	
Print Acceleration	
Travel Acceleration	
Print Jerk	
Travel Jerk	
Build Plate Adhesion Type	
Level of Support Structures	
Comments….	
Outputs	
Print Completion Time	
Smoothness Level (1-10)	
Weight	
Volume	
Noise Level	

Print Number _

Inputs	
Printer Name	
Print File Name	
Filament Material	
Bed Temperature	
Nozzle Temperature	
Nozzle Diameter	
Layer Height	
Initial Layer Height	
Flow	
Enable Retraction (Yes/No)	
Wall Thickness	
Travel Speed	
Print Acceleration	
Travel Acceleration	
Print Jerk	
Travel Jerk	
Build Plate Adhesion Type	
Level of Support Structures	
Comments....	
Outputs	
Print Completion Time	
Smoothness Level (1-10)	
Weight	
Volume	
Noise Level	

Print Number _

Inputs	
Printer Name	
Print File Name	
Filament Material	
Bed Temperature	
Nozzle Temperature	
Nozzle Diameter	
Layer Height	
Initial Layer Height	
Flow	
Enable Retraction (Yes/No)	
Wall Thickness	
Travel Speed	
Print Acceleration	
Travel Acceleration	
Print Jerk	
Travel Jerk	
Build Plate Adhesion Type	
Level of Support Structures	
Comments….	
Outputs	
Print Completion Time	
Smoothness Level (1-10)	
Weight	
Volume	
Noise Level	

Print Number _

Inputs	
Printer Name	
Print File Name	
Filament Material	
Bed Temperature	
Nozzle Temperature	
Nozzle Diameter	
Layer Height	
Initial Layer Height	
Flow	
Enable Retraction (Yes/No)	
Wall Thickness	
Travel Speed	
Print Acceleration	
Travel Acceleration	
Print Jerk	
Travel Jerk	
Build Plate Adhesion Type	
Level of Support Structures	
Comments....	
Outputs	
Print Completion Time	
Smoothness Level (1-10)	
Weight	
Volume	
Noise Level	

Print Number _

Inputs	
Printer Name	
Print File Name	
Filament Material	
Bed Temperature	
Nozzle Temperature	
Nozzle Diameter	
Layer Height	
Initial Layer Height	
Flow	
Enable Retraction (Yes/No)	
Wall Thickness	
Travel Speed	
Print Acceleration	
Travel Acceleration	
Print Jerk	
Travel Jerk	
Build Plate Adhesion Type	
Level of Support Structures	
Comments….	
Outputs	
Print Completion Time	
Smoothness Level (1-10)	
Weight	
Volume	
Noise Level	

Print Number _

Inputs	
Printer Name	
Print File Name	
Filament Material	
Bed Temperature	
Nozzle Temperature	
Nozzle Diameter	
Layer Height	
Initial Layer Height	
Flow	
Enable Retraction (Yes/No)	
Wall Thickness	
Travel Speed	
Print Acceleration	
Travel Acceleration	
Print Jerk	
Travel Jerk	
Build Plate Adhesion Type	
Level of Support Structures	
Comments….	
Outputs	
Print Completion Time	
Smoothness Level (1-10)	
Weight	
Volume	
Noise Level	

Print Number _

Inputs	
Printer Name	
Print File Name	
Filament Material	
Bed Temperature	
Nozzle Temperature	
Nozzle Diameter	
Layer Height	
Initial Layer Height	
Flow	
Enable Retraction (Yes/No)	
Wall Thickness	
Travel Speed	
Print Acceleration	
Travel Acceleration	
Print Jerk	
Travel Jerk	
Build Plate Adhesion Type	
Level of Support Structures	
Comments….	
Outputs	
Print Completion Time	
Smoothness Level (1-10)	
Weight	
Volume	
Noise Level	

Print Number _

Inputs	
Printer Name	
Print File Name	
Filament Material	
Bed Temperature	
Nozzle Temperature	
Nozzle Diameter	
Layer Height	
Initial Layer Height	
Flow	
Enable Retraction (Yes/No)	
Wall Thickness	
Travel Speed	
Print Acceleration	
Travel Acceleration	
Print Jerk	
Travel Jerk	
Build Plate Adhesion Type	
Level of Support Structures	
Comments....	
Outputs	
Print Completion Time	
Smoothness Level (1-10)	
Weight	
Volume	
Noise Level	

Print Number _

Inputs	
Printer Name	
Print File Name	
Filament Material	
Bed Temperature	
Nozzle Temperature	
Nozzle Diameter	
Layer Height	
Initial Layer Height	
Flow	
Enable Retraction (Yes/No)	
Wall Thickness	
Travel Speed	
Print Acceleration	
Travel Acceleration	
Print Jerk	
Travel Jerk	
Build Plate Adhesion Type	
Level of Support Structures	
Comments….	
Outputs	
Print Completion Time	
Smoothness Level (1-10)	
Weight	
Volume	
Noise Level	

Print Number _

Inputs	
Printer Name	
Print File Name	
Filament Material	
Bed Temperature	
Nozzle Temperature	
Nozzle Diameter	
Layer Height	
Initial Layer Height	
Flow	
Enable Retraction (Yes/No)	
Wall Thickness	
Travel Speed	
Print Acceleration	
Travel Acceleration	
Print Jerk	
Travel Jerk	
Build Plate Adhesion Type	
Level of Support Structures	
Comments….	
Outputs	
Print Completion Time	
Smoothness Level (1-10)	
Weight	
Volume	
Noise Level	

Print Number _

Inputs	
Printer Name	
Print File Name	
Filament Material	
Bed Temperature	
Nozzle Temperature	
Nozzle Diameter	
Layer Height	
Initial Layer Height	
Flow	
Enable Retraction (Yes/No)	
Wall Thickness	
Travel Speed	
Print Acceleration	
Travel Acceleration	
Print Jerk	
Travel Jerk	
Build Plate Adhesion Type	
Level of Support Structures	
Comments....	
Outputs	
Print Completion Time	
Smoothness Level (1-10)	
Weight	
Volume	
Noise Level	

Print Number _

Inputs	
Printer Name	
Print File Name	
Filament Material	
Bed Temperature	
Nozzle Temperature	
Nozzle Diameter	
Layer Height	
Initial Layer Height	
Flow	
Enable Retraction (Yes/No)	
Wall Thickness	
Travel Speed	
Print Acceleration	
Travel Acceleration	
Print Jerk	
Travel Jerk	
Build Plate Adhesion Type	
Level of Support Structures	
Comments….	
Outputs	
Print Completion Time	
Smoothness Level (1-10)	
Weight	
Volume	
Noise Level	

Print Number _

Inputs	
Printer Name	
Print File Name	
Filament Material	
Bed Temperature	
Nozzle Temperature	
Nozzle Diameter	
Layer Height	
Initial Layer Height	
Flow	
Enable Retraction (Yes/No)	
Wall Thickness	
Travel Speed	
Print Acceleration	
Travel Acceleration	
Print Jerk	
Travel Jerk	
Build Plate Adhesion Type	
Level of Support Structures	
Comments….	
Outputs	
Print Completion Time	
Smoothness Level (1-10)	
Weight	
Volume	
Noise Level	

Print Number _

Inputs	
Printer Name	
Print File Name	
Filament Material	
Bed Temperature	
Nozzle Temperature	
Nozzle Diameter	
Layer Height	
Initial Layer Height	
Flow	
Enable Retraction (Yes/No)	
Wall Thickness	
Travel Speed	
Print Acceleration	
Travel Acceleration	
Print Jerk	
Travel Jerk	
Build Plate Adhesion Type	
Level of Support Structures	
Comments….	
Outputs	
Print Completion Time	
Smoothness Level (1-10)	
Weight	
Volume	
Noise Level	

Print Number _

Inputs	
Printer Name	
Print File Name	
Filament Material	
Bed Temperature	
Nozzle Temperature	
Nozzle Diameter	
Layer Height	
Initial Layer Height	
Flow	
Enable Retraction (Yes/No)	
Wall Thickness	
Travel Speed	
Print Acceleration	
Travel Acceleration	
Print Jerk	
Travel Jerk	
Build Plate Adhesion Type	
Level of Support Structures	
Comments….	
Outputs	
Print Completion Time	
Smoothness Level (1-10)	
Weight	
Volume	
Noise Level	

Print Number _

Inputs	
Printer Name	
Print File Name	
Filament Material	
Bed Temperature	
Nozzle Temperature	
Nozzle Diameter	
Layer Height	
Initial Layer Height	
Flow	
Enable Retraction (Yes/No)	
Wall Thickness	
Travel Speed	
Print Acceleration	
Travel Acceleration	
Print Jerk	
Travel Jerk	
Build Plate Adhesion Type	
Level of Support Structures	
Comments....	
Outputs	
Print Completion Time	
Smoothness Level (1-10)	
Weight	
Volume	
Noise Level	

Print Number _

Inputs	
Printer Name	
Print File Name	
Filament Material	
Bed Temperature	
Nozzle Temperature	
Nozzle Diameter	
Layer Height	
Initial Layer Height	
Flow	
Enable Retraction (Yes/No)	
Wall Thickness	
Travel Speed	
Print Acceleration	
Travel Acceleration	
Print Jerk	
Travel Jerk	
Build Plate Adhesion Type	
Level of Support Structures	
Comments....	
Outputs	
Print Completion Time	
Smoothness Level (1-10)	
Weight	
Volume	
Noise Level	

Print Number _

Inputs	
Printer Name	
Print File Name	
Filament Material	
Bed Temperature	
Nozzle Temperature	
Nozzle Diameter	
Layer Height	
Initial Layer Height	
Flow	
Enable Retraction (Yes/No)	
Wall Thickness	
Travel Speed	
Print Acceleration	
Travel Acceleration	
Print Jerk	
Travel Jerk	
Build Plate Adhesion Type	
Level of Support Structures	
Comments….	
Outputs	
Print Completion Time	
Smoothness Level (1-10)	
Weight	
Volume	
Noise Level	

Print Number _

Inputs	
Printer Name	
Print File Name	
Filament Material	
Bed Temperature	
Nozzle Temperature	
Nozzle Diameter	
Layer Height	
Initial Layer Height	
Flow	
Enable Retraction (Yes/No)	
Wall Thickness	
Travel Speed	
Print Acceleration	
Travel Acceleration	
Print Jerk	
Travel Jerk	
Build Plate Adhesion Type	
Level of Support Structures	
Comments….	
Outputs	
Print Completion Time	
Smoothness Level (1-10)	
Weight	
Volume	
Noise Level	

Print Number _

Inputs	
Printer Name	
Print File Name	
Filament Material	
Bed Temperature	
Nozzle Temperature	
Nozzle Diameter	
Layer Height	
Initial Layer Height	
Flow	
Enable Retraction (Yes/No)	
Wall Thickness	
Travel Speed	
Print Acceleration	
Travel Acceleration	
Print Jerk	
Travel Jerk	
Build Plate Adhesion Type	
Level of Support Structures	
Comments....	
Outputs	
Print Completion Time	
Smoothness Level (1-10)	
Weight	
Volume	
Noise Level	

Print Number _

Inputs	
Printer Name	
Print File Name	
Filament Material	
Bed Temperature	
Nozzle Temperature	
Nozzle Diameter	
Layer Height	
Initial Layer Height	
Flow	
Enable Retraction (Yes/No)	
Wall Thickness	
Travel Speed	
Print Acceleration	
Travel Acceleration	
Print Jerk	
Travel Jerk	
Build Plate Adhesion Type	
Level of Support Structures	
Comments....	
Outputs	
Print Completion Time	
Smoothness Level (1-10)	
Weight	
Volume	
Noise Level	

Print Number _

Inputs	
Printer Name	
Print File Name	
Filament Material	
Bed Temperature	
Nozzle Temperature	
Nozzle Diameter	
Layer Height	
Initial Layer Height	
Flow	
Enable Retraction (Yes/No)	
Wall Thickness	
Travel Speed	
Print Acceleration	
Travel Acceleration	
Print Jerk	
Travel Jerk	
Build Plate Adhesion Type	
Level of Support Structures	
Comments….	
Outputs	
Print Completion Time	
Smoothness Level (1-10)	
Weight	
Volume	
Noise Level	

Print Number _

Inputs	
Printer Name	
Print File Name	
Filament Material	
Bed Temperature	
Nozzle Temperature	
Nozzle Diameter	
Layer Height	
Initial Layer Height	
Flow	
Enable Retraction (Yes/No)	
Wall Thickness	
Travel Speed	
Print Acceleration	
Travel Acceleration	
Print Jerk	
Travel Jerk	
Build Plate Adhesion Type	
Level of Support Structures	
Comments….	
Outputs	
Print Completion Time	
Smoothness Level (1-10)	
Weight	
Volume	
Noise Level	

Print Number _

Inputs	
Printer Name	
Print File Name	
Filament Material	
Bed Temperature	
Nozzle Temperature	
Nozzle Diameter	
Layer Height	
Initial Layer Height	
Flow	
Enable Retraction (Yes/No)	
Wall Thickness	
Travel Speed	
Print Acceleration	
Travel Acceleration	
Print Jerk	
Travel Jerk	
Build Plate Adhesion Type	
Level of Support Structures	
Comments….	
Outputs	
Print Completion Time	
Smoothness Level (1-10)	
Weight	
Volume	
Noise Level	

Print Number _

Inputs	
Printer Name	
Print File Name	
Filament Material	
Bed Temperature	
Nozzle Temperature	
Nozzle Diameter	
Layer Height	
Initial Layer Height	
Flow	
Enable Retraction (Yes/No)	
Wall Thickness	
Travel Speed	
Print Acceleration	
Travel Acceleration	
Print Jerk	
Travel Jerk	
Build Plate Adhesion Type	
Level of Support Structures	
Comments….	
Outputs	
Print Completion Time	
Smoothness Level (1-10)	
Weight	
Volume	
Noise Level	

Print Number _

Inputs	
Printer Name	
Print File Name	
Filament Material	
Bed Temperature	
Nozzle Temperature	
Nozzle Diameter	
Layer Height	
Initial Layer Height	
Flow	
Enable Retraction (Yes/No)	
Wall Thickness	
Travel Speed	
Print Acceleration	
Travel Acceleration	
Print Jerk	
Travel Jerk	
Build Plate Adhesion Type	
Level of Support Structures	
Comments….	
Outputs	
Print Completion Time	
Smoothness Level (1-10)	
Weight	
Volume	
Noise Level	

Print Number _

Inputs	
Printer Name	
Print File Name	
Filament Material	
Bed Temperature	
Nozzle Temperature	
Nozzle Diameter	
Layer Height	
Initial Layer Height	
Flow	
Enable Retraction (Yes/No)	
Wall Thickness	
Travel Speed	
Print Acceleration	
Travel Acceleration	
Print Jerk	
Travel Jerk	
Build Plate Adhesion Type	
Level of Support Structures	
Comments….	
Outputs	
Print Completion Time	
Smoothness Level (1-10)	
Weight	
Volume	
Noise Level	

Print Number _

Inputs	
Printer Name	
Print File Name	
Filament Material	
Bed Temperature	
Nozzle Temperature	
Nozzle Diameter	
Layer Height	
Initial Layer Height	
Flow	
Enable Retraction (Yes/No)	
Wall Thickness	
Travel Speed	
Print Acceleration	
Travel Acceleration	
Print Jerk	
Travel Jerk	
Build Plate Adhesion Type	
Level of Support Structures	
Comments….	
Outputs	
Print Completion Time	
Smoothness Level (1-10)	
Weight	
Volume	
Noise Level	

Print Number _

Inputs	
Printer Name	
Print File Name	
Filament Material	
Bed Temperature	
Nozzle Temperature	
Nozzle Diameter	
Layer Height	
Initial Layer Height	
Flow	
Enable Retraction (Yes/No)	
Wall Thickness	
Travel Speed	
Print Acceleration	
Travel Acceleration	
Print Jerk	
Travel Jerk	
Build Plate Adhesion Type	
Level of Support Structures	
Comments….	
Outputs	
Print Completion Time	
Smoothness Level (1-10)	
Weight	
Volume	
Noise Level	

Print Number _

Inputs	
Printer Name	
Print File Name	
Filament Material	
Bed Temperature	
Nozzle Temperature	
Nozzle Diameter	
Layer Height	
Initial Layer Height	
Flow	
Enable Retraction (Yes/No)	
Wall Thickness	
Travel Speed	
Print Acceleration	
Travel Acceleration	
Print Jerk	
Travel Jerk	
Build Plate Adhesion Type	
Level of Support Structures	
Comments….	
Outputs	
Print Completion Time	
Smoothness Level (1-10)	
Weight	
Volume	
Noise Level	

Print Number _

Inputs	
Printer Name	
Print File Name	
Filament Material	
Bed Temperature	
Nozzle Temperature	
Nozzle Diameter	
Layer Height	
Initial Layer Height	
Flow	
Enable Retraction (Yes/No)	
Wall Thickness	
Travel Speed	
Print Acceleration	
Travel Acceleration	
Print Jerk	
Travel Jerk	
Build Plate Adhesion Type	
Level of Support Structures	
Comments....	
Outputs	
Print Completion Time	
Smoothness Level (1-10)	
Weight	
Volume	
Noise Level	

Print Number _

Inputs	
Printer Name	
Print File Name	
Filament Material	
Bed Temperature	
Nozzle Temperature	
Nozzle Diameter	
Layer Height	
Initial Layer Height	
Flow	
Enable Retraction (Yes/No)	
Wall Thickness	
Travel Speed	
Print Acceleration	
Travel Acceleration	
Print Jerk	
Travel Jerk	
Build Plate Adhesion Type	
Level of Support Structures	
Comments....	
Outputs	
Print Completion Time	
Smoothness Level (1-10)	
Weight	
Volume	
Noise Level	

Print Number _

Inputs	
Printer Name	
Print File Name	
Filament Material	
Bed Temperature	
Nozzle Temperature	
Nozzle Diameter	
Layer Height	
Initial Layer Height	
Flow	
Enable Retraction (Yes/No)	
Wall Thickness	
Travel Speed	
Print Acceleration	
Travel Acceleration	
Print Jerk	
Travel Jerk	
Build Plate Adhesion Type	
Level of Support Structures	
Comments….	
Outputs	
Print Completion Time	
Smoothness Level (1-10)	
Weight	
Volume	
Noise Level	

Print Number _

Inputs	
Printer Name	
Print File Name	
Filament Material	
Bed Temperature	
Nozzle Temperature	
Nozzle Diameter	
Layer Height	
Initial Layer Height	
Flow	
Enable Retraction (Yes/No)	
Wall Thickness	
Travel Speed	
Print Acceleration	
Travel Acceleration	
Print Jerk	
Travel Jerk	
Build Plate Adhesion Type	
Level of Support Structures	
Comments....	
Outputs	
Print Completion Time	
Smoothness Level (1-10)	
Weight	
Volume	
Noise Level	

Print Number _

Inputs	
Printer Name	
Print File Name	
Filament Material	
Bed Temperature	
Nozzle Temperature	
Nozzle Diameter	
Layer Height	
Initial Layer Height	
Flow	
Enable Retraction (Yes/No)	
Wall Thickness	
Travel Speed	
Print Acceleration	
Travel Acceleration	
Print Jerk	
Travel Jerk	
Build Plate Adhesion Type	
Level of Support Structures	
Comments....	
Outputs	
Print Completion Time	
Smoothness Level (1-10)	
Weight	
Volume	
Noise Level	

Print Number _

Inputs	
Printer Name	
Print File Name	
Filament Material	
Bed Temperature	
Nozzle Temperature	
Nozzle Diameter	
Layer Height	
Initial Layer Height	
Flow	
Enable Retraction (Yes/No)	
Wall Thickness	
Travel Speed	
Print Acceleration	
Travel Acceleration	
Print Jerk	
Travel Jerk	
Build Plate Adhesion Type	
Level of Support Structures	
Comments….	
Outputs	
Print Completion Time	
Smoothness Level (1-10)	
Weight	
Volume	
Noise Level	

Print Number _

Inputs	
Printer Name	
Print File Name	
Filament Material	
Bed Temperature	
Nozzle Temperature	
Nozzle Diameter	
Layer Height	
Initial Layer Height	
Flow	
Enable Retraction (Yes/No)	
Wall Thickness	
Travel Speed	
Print Acceleration	
Travel Acceleration	
Print Jerk	
Travel Jerk	
Build Plate Adhesion Type	
Level of Support Structures	
Comments….	
Outputs	
Print Completion Time	
Smoothness Level (1-10)	
Weight	
Volume	
Noise Level	

Print Number _

Inputs	
Printer Name	
Print File Name	
Filament Material	
Bed Temperature	
Nozzle Temperature	
Nozzle Diameter	
Layer Height	
Initial Layer Height	
Flow	
Enable Retraction (Yes/No)	
Wall Thickness	
Travel Speed	
Print Acceleration	
Travel Acceleration	
Print Jerk	
Travel Jerk	
Build Plate Adhesion Type	
Level of Support Structures	
Comments....	
Outputs	
Print Completion Time	
Smoothness Level (1-10)	
Weight	
Volume	
Noise Level	

Print Number _

Inputs	
Printer Name	
Print File Name	
Filament Material	
Bed Temperature	
Nozzle Temperature	
Nozzle Diameter	
Layer Height	
Initial Layer Height	
Flow	
Enable Retraction (Yes/No)	
Wall Thickness	
Travel Speed	
Print Acceleration	
Travel Acceleration	
Print Jerk	
Travel Jerk	
Build Plate Adhesion Type	
Level of Support Structures	
Comments….	
Outputs	
Print Completion Time	
Smoothness Level (1-10)	
Weight	
Volume	
Noise Level	

Print Number _

Inputs	
Printer Name	
Print File Name	
Filament Material	
Bed Temperature	
Nozzle Temperature	
Nozzle Diameter	
Layer Height	
Initial Layer Height	
Flow	
Enable Retraction (Yes/No)	
Wall Thickness	
Travel Speed	
Print Acceleration	
Travel Acceleration	
Print Jerk	
Travel Jerk	
Build Plate Adhesion Type	
Level of Support Structures	
Comments….	
Outputs	
Print Completion Time	
Smoothness Level (1-10)	
Weight	
Volume	
Noise Level	

Print Number _

Inputs	
Printer Name	
Print File Name	
Filament Material	
Bed Temperature	
Nozzle Temperature	
Nozzle Diameter	
Layer Height	
Initial Layer Height	
Flow	
Enable Retraction (Yes/No)	
Wall Thickness	
Travel Speed	
Print Acceleration	
Travel Acceleration	
Print Jerk	
Travel Jerk	
Build Plate Adhesion Type	
Level of Support Structures	
Comments....	
Outputs	
Print Completion Time	
Smoothness Level (1-10)	
Weight	
Volume	
Noise Level	

Print Number _

Inputs	
Printer Name	
Print File Name	
Filament Material	
Bed Temperature	
Nozzle Temperature	
Nozzle Diameter	
Layer Height	
Initial Layer Height	
Flow	
Enable Retraction (Yes/No)	
Wall Thickness	
Travel Speed	
Print Acceleration	
Travel Acceleration	
Print Jerk	
Travel Jerk	
Build Plate Adhesion Type	
Level of Support Structures	
Comments….	
Outputs	
Print Completion Time	
Smoothness Level (1-10)	
Weight	
Volume	
Noise Level	

Print Number _

Inputs	
Printer Name	
Print File Name	
Filament Material	
Bed Temperature	
Nozzle Temperature	
Nozzle Diameter	
Layer Height	
Initial Layer Height	
Flow	
Enable Retraction (Yes/No)	
Wall Thickness	
Travel Speed	
Print Acceleration	
Travel Acceleration	
Print Jerk	
Travel Jerk	
Build Plate Adhesion Type	
Level of Support Structures	
Comments….	
Outputs	
Print Completion Time	
Smoothness Level (1-10)	
Weight	
Volume	
Noise Level	

Print Number _

Inputs	
Printer Name	
Print File Name	
Filament Material	
Bed Temperature	
Nozzle Temperature	
Nozzle Diameter	
Layer Height	
Initial Layer Height	
Flow	
Enable Retraction (Yes/No)	
Wall Thickness	
Travel Speed	
Print Acceleration	
Travel Acceleration	
Print Jerk	
Travel Jerk	
Build Plate Adhesion Type	
Level of Support Structures	
Comments….	
Outputs	
Print Completion Time	
Smoothness Level (1-10)	
Weight	
Volume	
Noise Level	

Print Number _

Inputs	
Printer Name	
Print File Name	
Filament Material	
Bed Temperature	
Nozzle Temperature	
Nozzle Diameter	
Layer Height	
Initial Layer Height	
Flow	
Enable Retraction (Yes/No)	
Wall Thickness	
Travel Speed	
Print Acceleration	
Travel Acceleration	
Print Jerk	
Travel Jerk	
Build Plate Adhesion Type	
Level of Support Structures	
Comments….	
Outputs	
Print Completion Time	
Smoothness Level (1-10)	
Weight	
Volume	
Noise Level	

Print Number _

Inputs	
Printer Name	
Print File Name	
Filament Material	
Bed Temperature	
Nozzle Temperature	
Nozzle Diameter	
Layer Height	
Initial Layer Height	
Flow	
Enable Retraction (Yes/No)	
Wall Thickness	
Travel Speed	
Print Acceleration	
Travel Acceleration	
Print Jerk	
Travel Jerk	
Build Plate Adhesion Type	
Level of Support Structures	
Comments….	
Outputs	
Print Completion Time	
Smoothness Level (1-10)	
Weight	
Volume	
Noise Level	

Print Number _ ˙

Inputs	
Printer Name	
Print File Name	
Filament Material	
Bed Temperature	
Nozzle Temperature	
Nozzle Diameter	
Layer Height	
Initial Layer Height	
Flow	
Enable Retraction (Yes/No)	
Wall Thickness	
Travel Speed	
Print Acceleration	
Travel Acceleration	
Print Jerk	
Travel Jerk	
Build Plate Adhesion Type	
Level of Support Structures	
Comments....	
Outputs	
Print Completion Time	
Smoothness Level (1-10)	
Weight	
Volume	
Noise Level	

Failure LOG

Failure #_

Issue Number	
Date of Occurrence	
Issue Detailed Description	
Root Cause (if applicable)	
Short Term Fix	
Long Term Preventive Measure	

(Example) Failure #_

Issue Number	103
Date of Occurrence	February 1, 2016
Issue Detailed Description	Warping of part (mouse.html) on LH side bottom corner
Root Cause (if applicable)	Bed is uneven on lower side, causing print to not stick on LH side
Short Term Fix	Adjust bed height on LH bottom corner
Long Term Preventive Measure	TBD

Failure #_

Issue Number	
Date of Occurrence	
Issue Detailed Description	
Root Cause (if applicable)	
Short Term Fix	
Long Term Preventive Measure	

(Example) Failure #_

Issue Number	
Date of Occurrence	
Issue Detailed Description	
Root Cause (if applicable)	
Short Term Fix	
Long Term Preventive Measure	

Failure #_

Issue Number	
Date of Occurrence	
Issue Detailed Description	
Root Cause (if applicable)	
Short Term Fix	
Long Term Preventive Measure	

(Example) Failure #_

Issue Number	
Date of Occurrence	
Issue Detailed Description	
Root Cause (if applicable)	
Short Term Fix	
Long Term Preventive Measure	

Failure #_

Issue Number	
Date of Occurrence	
Issue Detailed Description	
Root Cause (if applicable)	
Short Term Fix	
Long Term Preventive Measure	

(Example) Failure #_

Issue Number	
Date of Occurrence	
Issue Detailed Description	
Root Cause (if applicable)	
Short Term Fix	
Long Term Preventive Measure	

Failure #_

Issue Number	
Date of Occurrence	
Issue Detailed Description	
Root Cause (if applicable)	
Short Term Fix	
Long Term Preventive Measure	

(Example) Failure #_

Issue Number	
Date of Occurrence	
Issue Detailed Description	
Root Cause (if applicable)	
Short Term Fix	
Long Term Preventive Measure	

Failure #_

Issue Number	
Date of Occurrence	
Issue Detailed Description	
Root Cause (if applicable)	
Short Term Fix	
Long Term Preventive Measure	

(Example) Failure #_

Issue Number	
Date of Occurrence	
Issue Detailed Description	
Root Cause (if applicable)	
Short Term Fix	
Long Term Preventive Measure	

Failure #_

Issue Number	
Date of Occurrence	
Issue Detailed Description	
Root Cause (if applicable)	
Short Term Fix	
Long Term Preventive Measure	

(Example) Failure #_

Issue Number	
Date of Occurrence	
Issue Detailed Description	
Root Cause (if applicable)	
Short Term Fix	
Long Term Preventive Measure	

R&D and Testing Log

Copyrighted
Experiment #_

Constants	

Input Variables										
Variable name	Run 1	Run 2	Run 3	Run 4	Run 5	Run 6	Run 7	Run 8	Run 9	Run 10
Output Variable										
Completion Time										
Weight										
Volume										
Smoothness (1-10)										
Accuracy										

118

Copyrighted
Experiment

Constants	
Printer Name	Ultimaker
Print File Name	pencilholder.stl
Filament Material	PLA
Bed Temperature	50 C
Nozzle Temperature	250 C
Nozzle Diameter	12 mm
Layer Height	0.3 mm

Input Variables											
Variable name	Run 1	Run 2	Run 3	Run 4	Run 5	Run 6	Run 7	Run 8	Run 9	Run 10	
Travel Speed (mm/s)	10	20	30	40	50	60	70	80	90	100	
Output Variable											
Completion Time (min)	234	204	180	165	133	112	89	78	65	54	
Weight (gm)	34	34	34	34	34	34	34	34	34	34	
Volume (cm3)											
Smoothness (1-10)	7	7	6	9	9	9	7	6	5	4	
Accuracy (mm)											

Experiment #_

Constants	

Input Variables											
Variable name	Run 1	Run 2	Run 3	Run 4	Run 5	Run 6	Run 7	Run 8	Run 9	Run 10	
Output Variable											
Completion Time											
Weight											
Volume											
Smoothness (1-10)											
Accuracy											

Copyrighted
Experiment #_

Constants	

Input Variables											
Variable name	Run 1	Run 2	Run 3	Run 4	Run 5	Run 6	Run 7	Run 8	Run 9	Run 10	
Output Variable											
Completion Time											
Weight											
Volume											
Smoothness (1-10)											
Accuracy											

Copyrighted

Experiment #_

Constants	

Input Variables											
Variable name	Run 1	Run 2	Run 3	Run 4	Run 5	Run 6	Run 7	Run 8	Run 9	Run 10	
Output Variable											
Completion Time											
Weight											
Volume											
Smoothness (1-10)											
Accuracy											

123

Experiment #_

Constants	

Input Variables											
Variable name	Run 1	Run 2	Run 3	Run 4	Run 5	Run 6	Run 7	Run 8	Run 9	Run 10	
Output Variable											
Completion Time											
Weight											
Volume											
Smoothness (1-10)											
Accuracy											

Copyrighted
Experiment #_

Constants	

Input Variables										
Variable name	Run 1	Run 2	Run 3	Run 4	Run 5	Run 6	Run 7	Run 8	Run 9	Run 10
Output Variable										
Completion Time										
Weight										
Volume										
Smoothness (1-10)										
Accuracy										

Copyrighted
Experiment #_

Constants	

Input Variables											
Variable name	Run 1	Run 2	Run 3	Run 4	Run 5	Run 6	Run 7	Run 8	Run 9	Run 10	
Output Variable											
Completion Time											
Weight											
Volume											
Smoothness (1-10)											
Accuracy											

Copyrighted
Experiment #_

Constants	

Input Variables											
Variable name	Run 1	Run 2	Run 3	Run 4	Run 5	Run 6	Run 7	Run 8	Run 9	Run 10	
Output Variable											
Completion Time											
Weight											
Volume											
Smoothness (1-10)											
Accuracy											

Copyrighted

Experiment #_

Constants	

Input Variables											
Variable name	Run 1	Run 2	Run 3	Run 4	Run 5	Run 6	Run 7	Run 8	Run 9	Run 10	
Output Variable											
Completion Time											
Weight											
Volume											
Smoothness (1-10)											
Accuracy											

Copyrighted
Experiment #_

Constants	

Input Variables											
Variable name	Run 1	Run 2	Run 3	Run 4	Run 5	Run 6	Run 7	Run 8	Run 9	Run 10	
Output Variable											
Completion Time											
Weight											
Volume											
Smoothness (1-10)											
Accuracy											

Experiment #_

Constants	

Input Variables											
Variable name	Run 1	Run 2	Run 3	Run 4	Run 5	Run 6	Run 7	Run 8	Run 9	Run 10	
Output Variable											
Completion Time											
Weight											
Volume											
Smoothness (1-10)											
Accuracy											

Copyrighted
Experiment #_

Constants	

Input Variables											
Variable name	Run 1	Run 2	Run 3	Run 4	Run 5	Run 6	Run 7	Run 8	Run 9	Run 10	
Output Variable											
Completion Time											
Weight											
Volume											
Smoothness (1-10)											
Accuracy											

Experiment #_

Constants	

Input Variables											
Variable name	Run 1	Run 2	Run 3	Run 4	Run 5	Run 6	Run 7	Run 8	Run 9	Run 10	
Output Variable											
Completion Time											
Weight											
Volume											
Smoothness (1-10)											
Accuracy											

Maintenance Log

Maintenance Log #_

Part Modified/Fixed/Cleaned	
Date Completed	
Operation Performed on Part	
Cost	
Was Part Replaced (Y/N)?	
Next Date	

(Example) Maintenance Log #345_

Part Modified/Fixed/Cleaned	Nozzle
Date Completed	February 9, 2015
Operation Performed on Part	Replacement
Cost	$15
Was Part Replaced (Y/N)?	Y
Next Date	After 100 more Prints

Maintenance Log #_

Part Modified/Fixed/Cleaned	
Date Completed	
Operation Performed on Part	
Cost	
Was Part Replaced (Y/N)?	
Next Date	

(Example) Maintenance Log __

Part Modified/Fixed/Cleaned	
Date Completed	
Operation Performed on Part	
Cost	
Was Part Replaced (Y/N)?	
Next Date	

Copyrighted
Maintenance Log #_

Part Modified/Fixed/Cleaned	
Date Completed	
Operation Performed on Part	
Cost	
Was Part Replaced (Y/N)?	
Next Date	

(Example) Maintenance Log __

Part Modified/Fixed/Cleaned	
Date Completed	
Operation Performed on Part	
Cost	
Was Part Replaced (Y/N)?	
Next Date	

Maintenance Log #_

Part Modified/Fixed/Cleaned	
Date Completed	
Operation Performed on Part	
Cost	
Was Part Replaced (Y/N)?	
Next Date	

(Example) Maintenance Log __

Part Modified/Fixed/Cleaned	
Date Completed	
Operation Performed on Part	
Cost	
Was Part Replaced (Y/N)?	
Next Date	

Maintenance Log #_

Part Modified/Fixed/Cleaned	
Date Completed	
Operation Performed on Part	
Cost	
Was Part Replaced (Y/N)?	
Next Date	

(Example) Maintenance Log __

Part Modified/Fixed/Cleaned	
Date Completed	
Operation Performed on Part	
Cost	
Was Part Replaced (Y/N)?	
Next Date	

Maintenance Log #_

Part Modified/Fixed/Cleaned	
Date Completed	
Operation Performed on Part	
Cost	
Was Part Replaced (Y/N)?	
Next Date	

(Example) Maintenance Log __

Part Modified/Fixed/Cleaned	
Date Completed	
Operation Performed on Part	
Cost	
Was Part Replaced (Y/N)?	
Next Date	

Copyrighted
Maintenance Log #_

Part Modified/Fixed/Cleaned	
Date Completed	
Operation Performed on Part	
Cost	
Was Part Replaced (Y/N)?	
Next Date	

(Example) Maintenance Log __

Part Modified/Fixed/Cleaned	
Date Completed	
Operation Performed on Part	
Cost	
Was Part Replaced (Y/N)?	
Next Date	

Copyrighted
Maintenance Log #_

Part Modified/Fixed/Cleaned	
Date Completed	
Operation Performed on Part	
Cost	
Was Part Replaced (Y/N)?	
Next Date	

(Example) Maintenance Log __

Part Modified/Fixed/Cleaned	
Date Completed	
Operation Performed on Part	
Cost	
Was Part Replaced (Y/N)?	
Next Date	

Maintenance Log #_

Part Modified/Fixed/Cleaned	
Date Completed	
Operation Performed on Part	
Cost	
Was Part Replaced (Y/N)?	
Next Date	

(Example) Maintenance Log __

Part Modified/Fixed/Cleaned	
Date Completed	
Operation Performed on Part	
Cost	
Was Part Replaced (Y/N)?	
Next Date	

Maintenance Log #_

Part Modified/Fixed/Cleaned	
Date Completed	
Operation Performed on Part	
Cost	
Was Part Replaced (Y/N)?	
Next Date	

(Example) Maintenance Log __

Part Modified/Fixed/Cleaned	
Date Completed	
Operation Performed on Part	
Cost	
Was Part Replaced (Y/N)?	
Next Date	

Copyrighted

150

Made in the USA
Middletown, DE
02 June 2021

40851464R00084